AF270198

Parasaurolophus

by Grace Hansen

abdobooks.com

Published by Abdo Kids, a division of ABDO, P.O. Box 398166, Minneapolis, Minnesota 55439.
Copyright © 2021 by Abdo Consulting Group, Inc. International copyrights reserved in all countries.
No part of this book may be reproduced in any form without written permission from the publisher.
Abdo Kids Jumbo™ is a trademark and logo of Abdo Kids.

Printed in the United States of America, North Mankato, Minnesota.

102020

012021

 THIS BOOK CONTAINS
RECYCLED MATERIALS

Photo Credits: iStock, Shutterstock, Thinkstock, ©missbossy p.19/CC BY 2.0

Production Contributors: Teddy Borth, Jennie Forsberg, Grace Hansen
Design Contributors: Dorothy Toth, Pakou Moua

Library of Congress Control Number: 2019956495

Publisher's Cataloging-in-Publication Data

Names: Hansen, Grace, author.

Title: Parasaurolophus / by Grace Hansen

Description: Minneapolis, Minnesota : Abdo Kids, 2021 | Series: Dinosaurs | Includes online resources and
    index.

Identifiers: ISBN 9781098202453 (lib. bdg.) | ISBN 9781098203436 (ebook) | ISBN 9781098203924
    (Read-to-Me ebook)

Subjects: LCSH: Parasaurolophus--Juvenile literature. | Dinosaurs--Juvenile literature. | Herbivores--
    Juvenile literature. | Paleontology--Cretaceous--Juvenile literature. | Dinosaurs--Behavior--Juvenile
    literature.

Classification: DDC 567.90--dc23

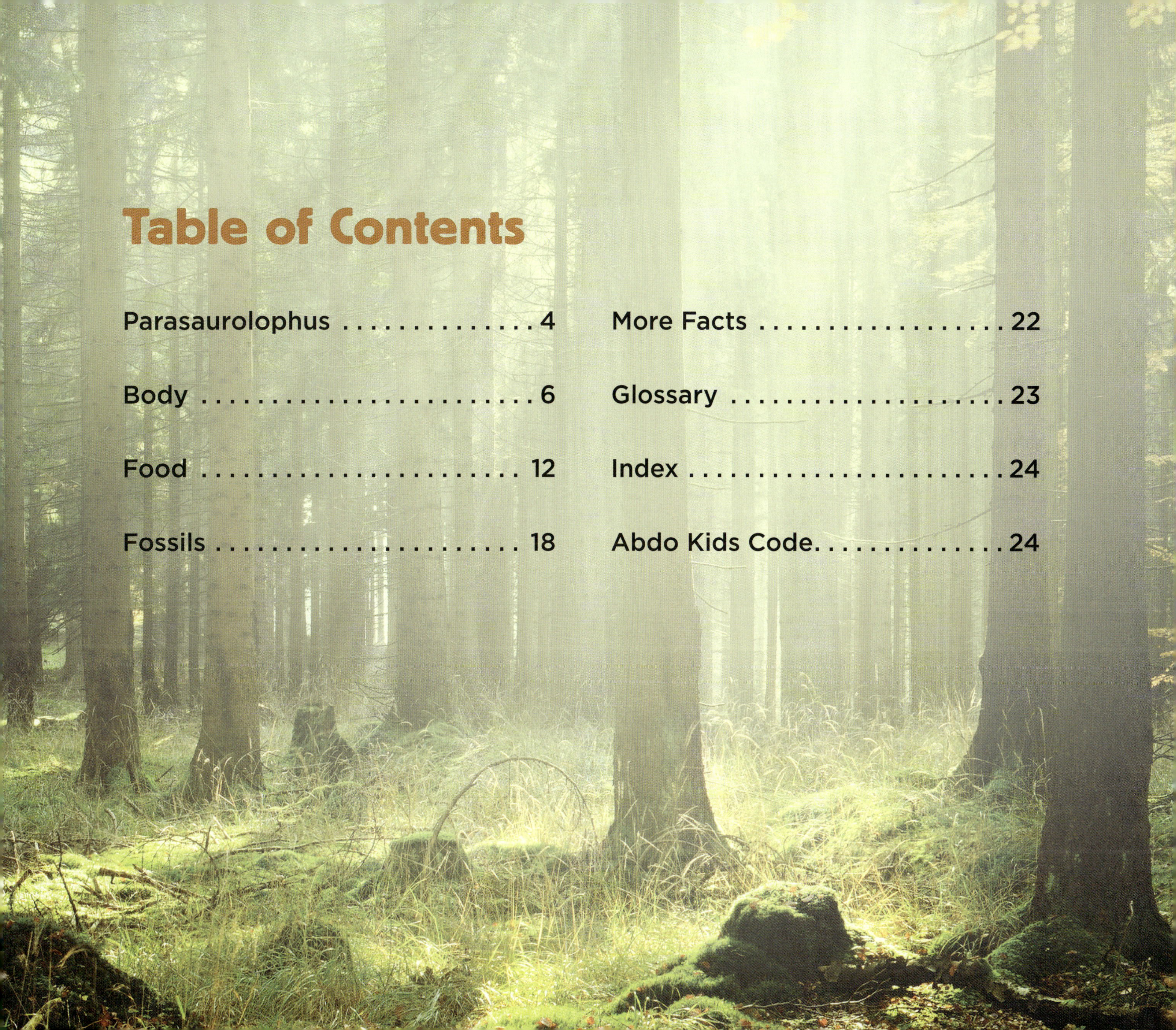

Table of Contents

Parasaurolophus

Parasaurolophus lived during the late **Cretaceous period**. That was around 75 million years ago. This dinosaur roamed land that is now North America.

4

Body

Parasaurolophus was about 32 feet (10 m) long from head to tail. It weighed around 8,000 pounds (3,628 kg).

Parasaurolophus (PAIR-uh-sor-AH-luh-fus)
is an ornithopod dinosaur
"Duck-billed" hadrosaurs had beaks, not bills
Parasaurolophus is a hadrosaur, a diverse group of ornithopods commonly known as "duck-billed" dinosaurs. But the nickname is misleading. The hadrosaur "bill" was nothing like a duck's toothless bill. At the front, a horny beak cropped bites of tough plant material. Hundreds of teeth then ground the leafy bite to a pulp.
The fossil in front o
The Field Museum
is a holotype, the
whether another
of the same species
to be identified as
cyrtocristatus, it
skeleton in its

Its head alone was around 6 feet (1.8 m) long. This included a large **crest** that the dinosaur is known for.

9

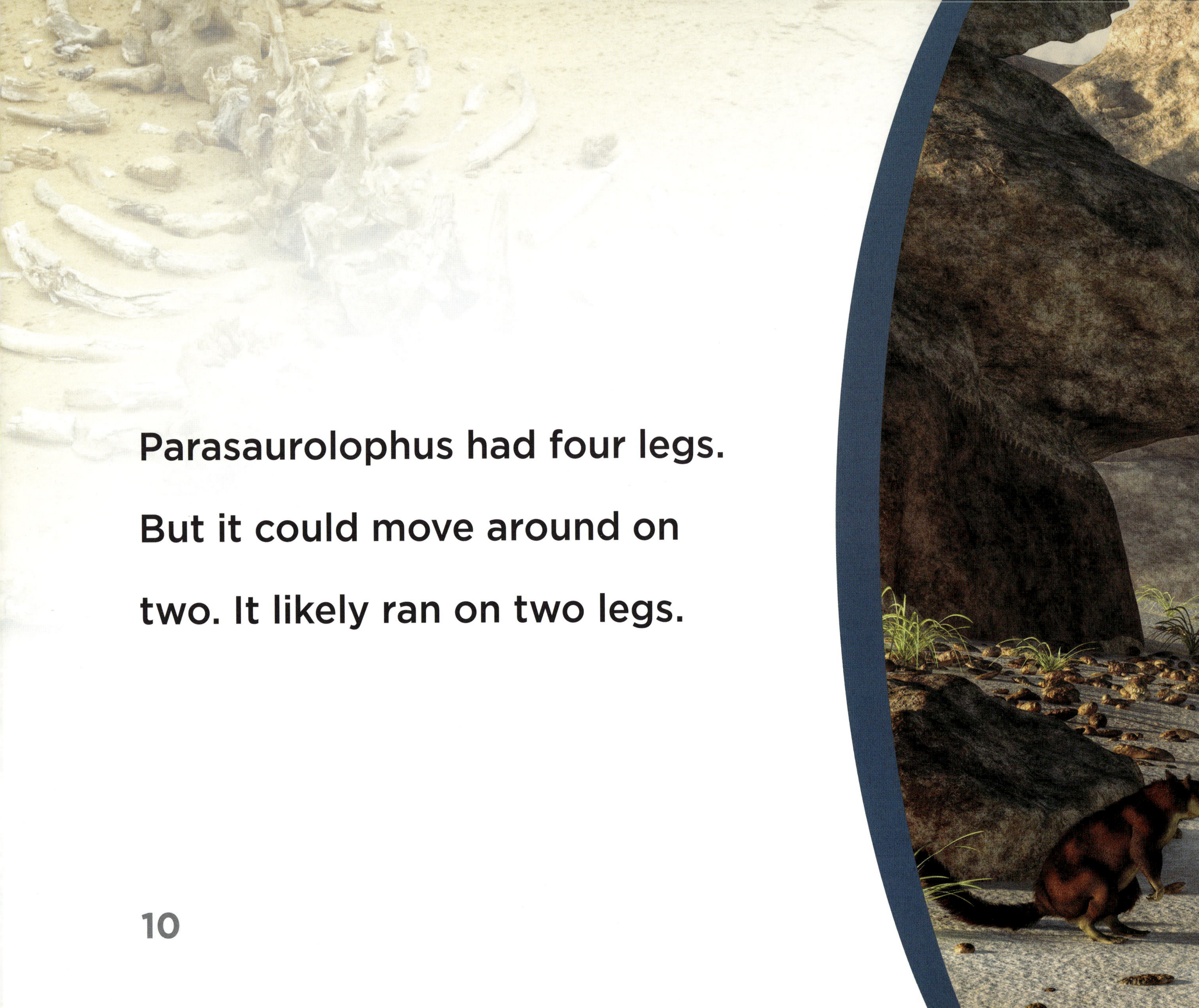

Parasaurolophus had four legs.
But it could move around on
two. It likely ran on two legs.

Food

Parasaurolophus was a **herbivore**. This means that it only ate plants.

It plucked plants with its bill-like mouth. It chewed up the plants in a grinding motion.

The grinding would wear down the dinosaur's teeth. New teeth would replace old ones.

Fossils

William Parks was a Canadian **paleontologist**. In 1920, he found the first Parasaurolophus **fossil** in Alberta.

Canada
Alberta
United States
William Parks
19

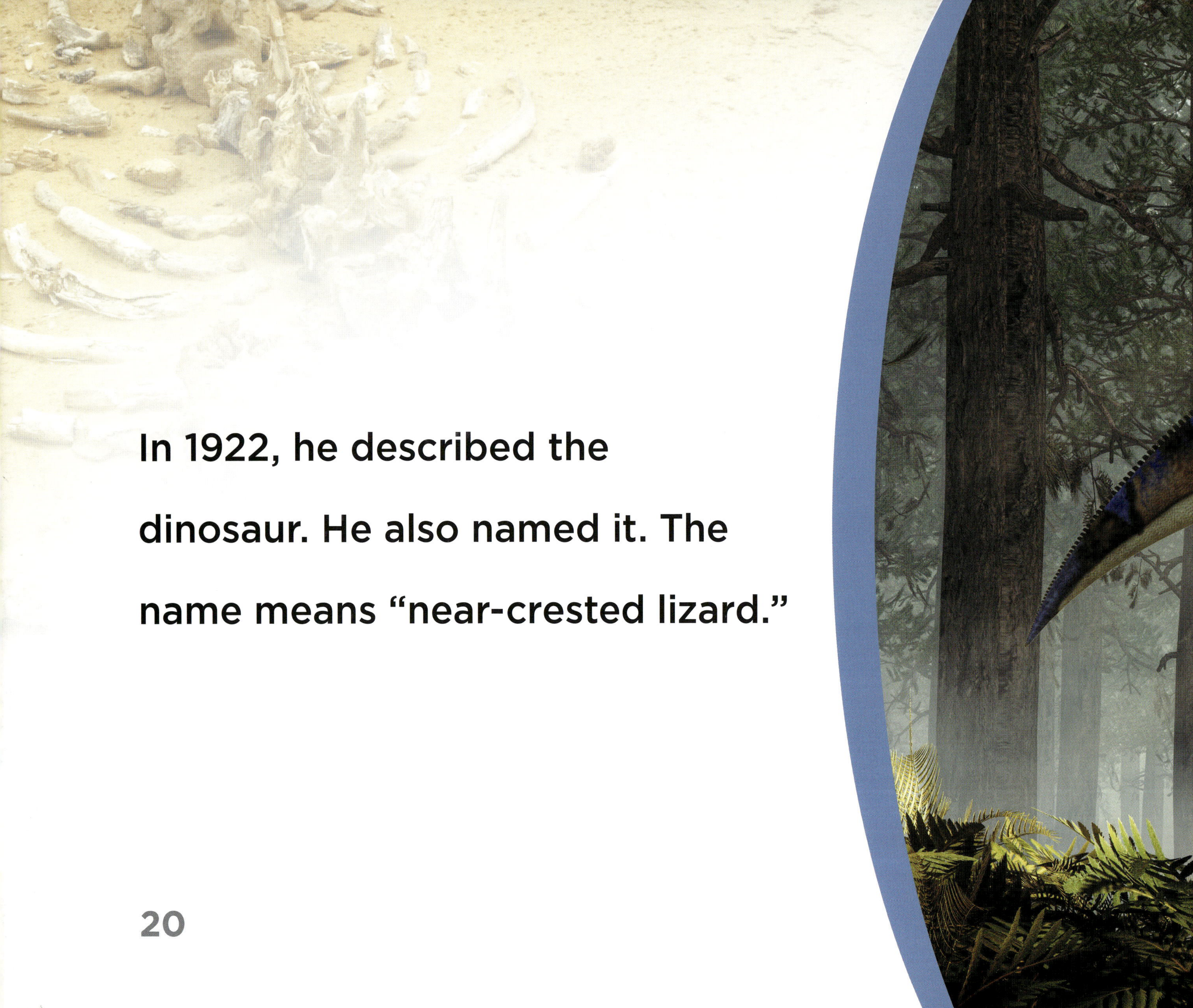

In 1922, he described the dinosaur. He also named it. The name means "near-crested lizard."

More Facts

- Parasaurolophus **fossils** have also been found in Utah and New Mexico.

- Scientists think the dinosaur might have used its **crest** to make a loud sound. It would use this sound to talk to other Parasaurolophus.

- This dinosaur was covered in small scales.

Glossary

crest – a tuft of feathers or bone on an animal's head.

Cretaceous period – a period of geological time that began 145 million years ago. The end of the Cretaceous period, about 66 million years ago, brought the mass extinction of the dinosaurs.

fossil – the remains, impression, or trace of something that lived long ago, as a skeleton, footprint, etc.

herbivore – an animal that only feeds on plants.

paleontologist – a scientist that studies paleontology, the science that studies animal and plant fossils for information about life in the past.

Index

Abdo Kids
ONLINE
FREE! ONLINE MULTIMEDIA RESOURCES

Visit **abdokids.com** to access crafts, games, videos, and more!

Use Abdo Kids code
DPK2453
or scan this QR code!